I0143170

# AI Lesson Planning Playbook

*100 Ready-to-Use Prompts to Save Time, Engage Students, and Differentiate Learning*

# Table of Contents

**Introduction** ..................................................5

    **Why Use AI for Lesson Planning** ................................6

    **How to Use This Playbook Effectively** ...........................6

    **Ethical & Responsible AI in the Classroom** ...................7

    **Tips for Adapting Prompts for Your Students**..................7

**Chapter 1 – Quick Lesson Outlines**...............................9

    **Purpose of This Chapter** ......................................9

    **Prompts**........................................................9

    **Pro Tip:**.......................................................11

    **Purpose of This Chapter** .....................................12

    **Prompts**.......................................................12

    **Pro Tip:**.......................................................14

**Chapter 3 – Activities & Engagement Prompts** ............16

    **Purpose of This Chapter** .....................................16

    **Prompts**.......................................................16

    **Pro Tip:**.......................................................18

**Chapter 4 – Assessment & Feedback Prompts**..............18

    **Purpose of This Chapter** .....................................19

    **Prompts**.......................................................19

Pro Tip:..............................................................................................21

## Chapter 5 – Math Prompts ...........................................................22

Purpose of This Chapter ............................................................22

Prompts.....................................................................................22

Pro Tip:.....................................................................................24

## Chapter 6 – Science Prompts .......................................................25

Purpose of This Chapter ............................................................25

Prompts.....................................................................................25

Pro Tip:.....................................................................................27

## Chapter 7 – English Language Arts (ELA) Prompts.....................28

Purpose of This Chapter ............................................................28

Prompts.....................................................................................28

Pro Tip:.....................................................................................30

## Chapter 8 – Social Studies Prompts .............................................31

Purpose of This Chapter ............................................................31

Prompts.....................................................................................31

Pro Tip:.....................................................................................33

## Chapter 9 – Arts, Music, and PE Prompts ...................................34

Purpose of This Chapter ............................................................34

Prompts.....................................................................................34

Pro Tip:.....................................................................................36

## Chapter 10 – Sub Plans with AI ...................................................37

Purpose of This Chapter ............................................................37

Prompts.....................................................................................37

Pro Tip:.....................................................................................39

## Chapter 11 – Flipped Classroom & Hybrid Learning Prompts ...40

Purpose of This Chapter ............................................................40

Prompts.....................................................................................40

Pro Tip:..................................................................42

**Chapter 12 – Project-Based Learning Units**..............43

Purpose of This Chapter ...................................43

Prompts...............................................................43

Pro Tip:..............................................................45

**Chapter 13 – Seasonal & Event-Based Lessons** .........46

Purpose of This Chapter ...................................46

Prompts...............................................................46

Pro Tip:..............................................................48

**Chapter 14 – Editable Prompt Templates** .................49

Purpose of This Chapter ...................................49

Templates ...........................................................49

Pro Tip:..............................................................51

**Chapter 15 – Prompt Adaptation Checklist**...............52

Purpose of This Chapter ...................................52

Step 1 – Clarify Your Goal................................52

Step 2 – Add Context.........................................52

Step 3 – Set Output Structure...........................52

Step 4 – Adjust for Learners.............................53

Step 5 – Build in Engagement ..........................53

Step 6 – Double-Check Accuracy .....................53

Step 7 – Make It Inclusive.................................53

Step 8 – Test and Tweak.....................................54

Pro Tip:..............................................................54

**Conclusion** ................................................................55

**Additional Resources**...............................................56

# Introduction

## Why Use AI for Lesson Planning

Lesson planning is essential, but it can also be one of the most time-consuming parts of teaching. Between aligning lessons to standards, differentiating for diverse learners, and finding engaging activities, many educators spend hours outside of class preparing.

Artificial Intelligence (AI) can help change that. With the right prompts, AI tools can generate lesson outlines, activities, assessments, and even differentiated resources in minutes. This doesn't replace your expertise—it enhances it. By letting AI handle the heavy lifting of brainstorming and structuring, you can focus on personalizing lessons, building relationships, and responding to your students' needs in real time.

AI is not just about saving time—it's about opening up more opportunities for creativity, exploration, and innovation in the classroom.

## How to Use This Playbook Effectively

This playbook is designed to be a **ready-to-use resource**. Each chapter contains AI prompts grouped by purpose—whether you need a quick lesson outline, a subject-specific activity, or a project-based learning unit.

Here's how to get the most from it:

1. **Choose Your Purpose** – Start with the section that matches your immediate need.
2. **Fill in the Details** – Every prompt has placeholders (e.g., [subject], [grade level], [topic]) for you to customize.

3. **Run the Prompt** – Paste your completed prompt into your AI tool of choice (e.g., ChatGPT, MagicSchool.ai, Curipod).
4. **Review & Adapt** – Always check for accuracy, alignment with standards, and cultural relevance before using the AI-generated material in your classroom.

Keep in mind—AI outputs are starting points, not final products. Think of them as a draft you can shape to your unique teaching style.

## Ethical & Responsible AI in the Classroom

As educators, we bear the responsibility of modeling ethical technology use for our students. This means:

- **Protecting Student Privacy** – Avoid entering any personal student data into AI tools.
- **Verifying Accuracy** – Double-check facts, figures, and historical information.
- **Avoiding Bias** – AI can reflect biases in its training data. Review outputs to ensure inclusivity and cultural sensitivity.
- **Transparency** – When appropriate, let students know when you've used AI in lesson preparation to promote openness about technology use.

By using AI responsibly, we not only enhance our workflow but also teach students valuable lessons about critical thinking and digital citizenship.

## Tips for Adapting Prompts for Your Students

The real power of AI comes from personalization. Here's how to adapt prompts for your classroom:

- **Specify the Reading Level** – This ensures materials are accessible for your students.
- **Incorporate Local Context** – Reference community events, local history, or familiar examples.
- **Adjust for Learning Styles** – Ask AI to produce visual aids, hands-on activities, or discussion-based tasks based on your students' needs.
- **Differentiate for Abilities** – Request multiple versions of an activity for varying skill levels.
- **Add Student Choice** – Include prompts that give students options in how they demonstrate their understanding.

When you adapt prompts, you're combining the speed of AI with the insight of an experienced educator—and that's where the magic happens.

# Chapter 1 – Quick Lesson Outlines

## Purpose of This Chapter

Quick lesson outlines are one of the fastest ways to save planning time with AI. By giving a structured, specific prompt, you can generate a standards-aligned, engaging lesson framework in seconds, then adapt it to your students' needs.

## Prompts

### 1. General Lesson Outline Prompt

Create a 45-minute [subject] lesson for [grade level] on [topic]. Include:

1. A learning objective aligned with [specific state or national standard]
2. An engaging opening activity
3. Two main learning activities
4. One formative assessment
5. Homework or extension activity

### 2. Cross-Curricular Lesson Prompt

Plan a 60-minute lesson for [grade level] that connects [subject] with [second subject]. Focus on the topic of [topic] and include at least one hands-on activity that reinforces both subjects.

### 3. Warm-Up Focus Prompt

Create a 10-minute warm-up activity to introduce [topic] for [grade level]. Include an engaging question, a quick group activity, and a connection to real-world examples.

---

### 4. Standards-Driven Outline Prompt

Design a [subject] lesson for [grade level] on [topic] aligned with [specific standard code]. Provide a clear objective, an engaging introduction, two learning activities, and a quick formative check for understanding.

---

### 5. Lesson with Differentiation Prompt

Create a 50-minute [subject] lesson on [topic] for [grade level]. Include:

- One activity for struggling learners
- One activity for advanced learners
- One activity for mixed-ability groups

---

### 6. Inquiry-Based Lesson Prompt

Develop a 45-minute inquiry-based lesson for [grade level] in [subject] on [topic]. Include guiding questions, exploration activities, and a short reflection prompt for students.

---

### 7. Group Work Emphasis Prompt

Plan a [subject] lesson for [grade level] on [topic] that relies heavily on group work. Include a team-building warm-up, a group project outline, and peer review steps.

## 8. Real-World Connection Prompt

Design a [subject] lesson for [grade level] on [topic] that connects to a real-world issue. Include an introduction that explains the relevance, a case study, and an activity where students propose solutions.

## 9. Technology-Integrated Lesson Prompt

Create a 50-minute [subject] lesson for [grade level] on [topic] that uses [specific digital tool] as part of the learning process. Include a tutorial segment, student activity, and assessment idea.

## 10. Review/Recap Lesson Prompt

Plan a review lesson for [grade level] in [subject] covering [unit or topic]. Include a brief review activity, a student-led recap, and a short quiz or game to check understanding.

# Pro Tip:

When using these prompts, always specify **subject, grade level,** and **learning goals** to get the best output. If your state or district has specific curriculum frameworks, include them in the prompt for better alignment.

# Chapter 2 – Differentiated Learning Prompts

## Purpose of This Chapter

Differentiation ensures that all students can access the duplicate content in ways that match their readiness, interests, and learning preferences. These prompts are designed to help teachers quickly generate multiple versions of a lesson or activity that cater to the diverse needs of learners.

---

## Prompts

### 1. Tiered Activity Prompt

Create three versions of an activity for [subject] on [topic] for [grade level]:

- Level 1: Basic understanding
- Level 2: Intermediate challenge
- Level 3: Advanced application

---

### 2. Multiple Reading Levels Prompt

Provide a reading passage on [topic] for [grade level] at three different reading levels. Include comprehension questions for each level.

---

### 3. Language Support Prompt

Adapt a [subject] lesson on [topic] for English learners at a [beginner/intermediate/advanced] proficiency level. Include vocabulary lists, sentence frames, and visual aids.

---

## 4. Learning Style Adaptation Prompt

Design three versions of a 45-minute [subject] lesson on [topic]:

- Visual learner version
- Auditory learner version
- Kinesthetic learner version

---

## 5. Choice Board Prompt

Create a choice board for [topic] in [subject] for [grade level] with at least 6 different activities that appeal to different interests and learning preferences.

---

## 6. Support & Enrichment Prompt

Develop a lesson plan on [topic] for [grade level] in [subject] that includes:

- A scaffolded activity for struggling learners
- An enrichment task for advanced learners

---

## 7. Vocabulary Scaffolding Prompt

Generate a vocabulary-building activity for [topic] in [subject] for [grade level]. Include a basic version for struggling readers and an advanced version for fluent readers.

---

### 8. Flexible Grouping Prompt

Plan a [subject] lesson on [topic] for [grade level] that includes activities for:

- Homogeneous skill groups
- Heterogeneous mixed-ability groups

---

### 9. Task Complexity Prompt

Create three versions of a [subject] project on [topic]:

- Short, guided project with clear steps
- Medium-level project with partial guidance
- Open-ended project requiring independent research

---

### 10. Multilingual Support Prompt

Provide an activity on [topic] for [subject] at [grade level] in both English and [second language]. Include key vocabulary, instructions, and sample answers in both languages.

---

## Pro Tip:

If you use a prompt like "Create three versions…" with an AI tool, clearly label the levels you want and specify the skill differences.

This makes the AI's output much easier to use without heavy editing.

# Chapter 3 – Activities & Engagement Prompts

## Purpose of This Chapter

Engaged students learn better. These prompts are designed to help teachers quickly create activities that encourage participation, curiosity, and collaboration, whether in-person or online.

## Prompts

### 1. Gamified Learning Prompt

Create a game-based activity for [subject] on [topic] for [grade level]. Include rules, scoring, and how to adapt it for groups or individuals.

### 2. Roleplay Scenario Prompt

Develop a roleplay activity for [subject] on [topic] for [grade level]. Assign student roles, provide a scenario, and list discussion or decision-making steps.

### 3. Mystery/Challenge Prompt

Plan a mystery or challenge activity for [subject] on [topic] for [grade level]. Include clues, problem-solving steps, and a solution reveal.

### 4. Hands-On Experiment Prompt

Design a hands-on experiment for [topic] in [subject] for [grade level]. Include materials, step-by-step instructions, safety notes, and a reflection question.

---

### 5. Creative Project Prompt

Create a project-based assignment for [subject] on [topic] for [grade level] that results in a tangible product (poster, video, model, presentation). Include rubrics and timeline suggestions.

---

### 6. Debate Activity Prompt

Develop a debate lesson for [subject] on [controversial or discussion-worthy topic] for [grade level]. Include a debate format, roles, sample arguments, and reflection questions.

---

### 7. Interactive Digital Activity Prompt

Create a digital scavenger hunt for [subject] on [topic] for [grade level] using free online tools. Include instructions, link ideas, and a scoring method.

---

### 8. Peer Teaching Prompt

Plan a peer-teaching activity where students learn about [topic] in [subject] and then teach it to classmates. Include preparation steps, teaching tips, and peer feedback forms.

### 9. Artistic Expression Prompt

Develop an activity for [subject] on [topic] for [grade level] that uses art, music, or drama to express understanding. Include clear expectations and evaluation criteria.

### 10. Simulation Prompt

Design a simulation for [subject] on [topic] for [grade level] that mimics a real-life event or process. Include setup, roles, and debrief discussion points.

## Pro Tip:

When creating engagement activities with AI, always specify the **time limit, group size**, and **available resources**. This keeps outputs realistic and classroom-ready.

# Chapter 4 – Assessment & Feedback Prompts

## Purpose of This Chapter

Assessment and feedback help guide student learning, but they can be time-consuming. These prompts help teachers quickly create effective quizzes, rubrics, and feedback that are aligned with objectives and easy to adapt.

## Prompts

### 1. Quick Quiz Prompt

Create a 10-question multiple-choice quiz for [subject] on [topic] for [grade level]. Include an answer key and explanations for each correct answer.

### 2. Formative Assessment Prompt

Design three short formative assessment activities for [subject] on [topic] for [grade level] that can be completed in under 10 minutes.

### 3. Exit Ticket Prompt

Generate five exit ticket questions for [subject] on [topic] for [grade level] that check for understanding and encourage reflection.

## 4. Open-Ended Question Prompt

Create five open-ended discussion or reflection questions for [subject] on [topic] for [grade level] to encourage critical thinking.

---

## 5. Project Rubric Prompt

Develop a grading rubric for a [subject] project on [topic] for [grade level]. Include 4–5 criteria, each with descriptions for Excellent, Good, Fair, and Needs Improvement.

---

## 6. Peer Assessment Prompt

Create a peer feedback form for [subject] presentations or projects. Include rating scales, constructive feedback prompts, and space for positive comments.

---

## 7. Self-Assessment Prompt

Design a student self-assessment checklist for [subject] on [topic] for [grade level] that encourages reflection on effort, understanding, and participation.

---

## 8. Quick Feedback Statement Prompt

Generate 10 feedback comments for [subject] on [topic] for [grade level] that highlight strengths, suggest improvements, and encourage growth.

---

### 9. Misconception Identification Prompt

List common misconceptions students might have about [topic] in [subject] for [grade level], and suggest short activities to address each one.

---

### 10. Standards-Based Assessment Prompt

Create an assessment for [subject] on [topic] for [grade level] that directly measures skills in [specific state/national standard]. Include scoring guidelines.

---

# Pro Tip:

When generating assessments with AI, always **review them for accuracy, ensure that the difficulty matches your students' abilities, and adapt the wording to align with** your classroom culture and tone.

# Chapter 5 – Math Prompts

## Purpose of This Chapter

Math teachers often need a variety of practice problems, engaging activities, and clear explanations tailored to different learning levels. These prompts help you create AI-generated math resources that are accurate, aligned with standards, and adaptable to your class.

## Prompts

### 1. Standards-Aligned Lesson Prompt

Create a 45-minute [grade level] math lesson on [topic] aligned with [specific standard]. Include learning objectives, example problems, guided practice, and an exit ticket.

### 2. Differentiated Problem Sets Prompt

Generate 15 math problems on [topic] for [grade level]:

- 5 easy
- 5 medium
- 5 challenging
  Include answer keys and explanations for each.

### 3. Real-World Application Prompt

Design a math activity for [grade level] on [topic] that applies the concept to a real-world scenario. Include clear instructions, example calculations, and reflection questions.

## 4. Math Game Prompt

Create a math game for [topic] at [grade level] that can be played in small groups. Include rules, scoring, and sample problems.

## 5. Visual Explanation Prompt

Provide a visual explanation of [topic] for [grade level], including diagrams or step-by-step illustrations and a written description.

## 6. Word Problem Generator Prompt

Generate 10-word problems on [topic] for [grade level]. Include a variety of real-life contexts and provide complete solutions.

## 7. Common Mistakes Prompt

List the 5 most common mistakes students make with [topic] in [grade level] math and provide example problems showing the correct method.

## 8. Math Vocabulary Prompt

Create a vocabulary list for [topic] in [grade level] math, including definitions, examples, and illustrations if applicable.

### 9. Cross-Disciplinary Prompt

Design a math activity for [topic] that connects with [science, art, or technology]. Include step-by-step instructions and discussion questions.

---

### 10. Challenge Problem Prompt

Create three challenge problems on [topic] for advanced [grade level] math students. Include detailed solutions and explanations of the reasoning process.

---

## Pro Tip:

When working with AI for math content, always **double-check calculations**. Even strong AI tools can make math errors, so verify before giving problems to students.

# Chapter 6 – Science Prompts

## Purpose of This Chapter

Science instruction thrives on curiosity, experimentation, and real-world connections. These prompts help teachers generate AI-created lessons, labs, and STEM activities that are aligned with standards and accessible for all learners.

## Prompts

### 1. Standards-Aligned Science Lesson Prompt

Create a 45-minute [grade level] science lesson on [topic] aligned with [specific standard]. Include objectives, a hands-on activity, and an exit ticket question.

### 2. Inquiry-Based Activity Prompt

Design an inquiry-based investigation for [topic] in [science branch: biology, chemistry, physics, earth science] for [grade level]. Include guiding questions, materials, and procedure steps.

### 3. Lab Experiment Prompt

Generate a lab activity for [topic] for [grade level]. Include safety notes, materials list, step-by-step instructions, and follow-up reflection questions.

### 4. Science Vocabulary Prompt

Create a vocabulary list for [topic] in [grade level] science. Include definitions, visual examples, and a short quiz.

---

## 5. Real-World Application Prompt

Plan a science activity for [topic] in [grade level] that connects to a current event or environmental issue. Include background information, discussion questions, and a solution-focused task.

---

## 6. STEM/STEAM Integration Prompt

Design a STEM/STEAM project that teaches [topic] in [grade level] science and incorporates [technology, engineering, or art]. Include project guidelines and assessment criteria.

---

## 7. Scientific Method Practice Prompt

Create a [grade level] science lesson on [topic] that walks students through each step of the scientific method, using an age-appropriate experiment.

---

## 8. Model/Simulation Prompt

Generate instructions for building a model or simulation to demonstrate [topic] for [grade level]. Include materials, assembly steps, and discussion questions.

---

## 9. Common Misconceptions Prompt

List common misconceptions students have about [topic] in [grade level] science, and create short activities to correct them.

## 10. Cross-Curricular Science Prompt

Create a lesson on [topic] that connects science with [math, ELA, or social studies]. Include an integrated activity and reflection component.

## Pro Tip:

When generating lab activities or experiments with AI, **double-check safety instructions** and adapt materials to what's realistically available in your school.

# Chapter 7 – English Language Arts (ELA) Prompts

## Purpose of This Chapter

ELA lessons benefit from variety, creativity, and opportunities for students to express themselves. These prompts enable teachers to quickly create AI-generated reading comprehension tasks, writing exercises, and vocabulary activities that are adaptable for various grades and abilities.

## Prompts

### 1. Reading Comprehension Prompt

Create a reading comprehension passage for [grade level] on [topic or theme]. Include 5 multiple-choice questions, 2 short-answer questions, and an answer key.

### 2. Literary Analysis Prompt

Generate a set of 5 literary analysis questions for [grade level] students based on [text or excerpt]. Include a short teacher's guide for expected responses.

### 3. Creative Writing Starter Prompt

Provide 5 creative writing prompts for [grade level] based on [theme or setting]. Include one sentence starter and one image prompt for each.

## 4. Grammar Practice Prompt

Create 10 grammar exercises for [grade level] focused on [specific grammar concept], with answer key and explanations.

## 5. Vocabulary Expansion Prompt

Generate a vocabulary list of 15 words related to [theme or text] for [grade level]. Include definitions, sample sentences, and a short vocabulary quiz.

## 6. Debate and Discussion Prompt

Create a set of discussion questions for [grade level] based on [text, theme, or current event]. Include one controversial or open-ended question to spark debate.

## 7. Poetry Exploration Prompt

Plan a poetry lesson for [grade level] on [poetic form or theme]. Include a mentor text, analysis questions, and a guided writing activity.

## 8. Compare & Contrast Prompt

Create an activity for [grade level] where students compare and contrast two [characters, settings, or themes] from [text(s)]. Include guiding questions and a graphic organizer.

### 9. Real-World Connection Prompt

Develop a writing activity for [grade level] that connects a theme from [text] to a real-world issue. Include a short research component and discussion questions.

---

### 10. Public Speaking Prompt

Design a public speaking activity for [grade level] where students present a summary or analysis of [text or theme]. Include guidelines for structure, delivery tips, and peer feedback criteria.

---

## Pro Tip:

When using AI for ELA content, specify **reading level, genre, and tone** to get outputs that match your students' abilities and interests.

# Chapter 8 – Social Studies Prompts

## Purpose of This Chapter

Social Studies brings together history, geography, government, and culture. These prompts enable teachers to create AI-generated lessons that connect the past to the present, encourage critical thinking, and make the subject more relevant for students.

## Prompts

### 1. Historical Event Lesson Prompt

Create a 45-minute [grade level] lesson on [historical event]. Include a brief background, 2 engaging activities, and a short formative assessment.

### 2. Primary Source Analysis Prompt

Provide a classroom activity for [grade level] where students analyze a primary source related to [topic]. Include guiding questions and a short writing assignment.

### 3. Geography Exploration Prompt

Design an interactive geography activity for [grade level] on [region or concept]. Include a map-based task, discussion questions, and a short quiz.

### 4. Civics & Government Prompt

Create a lesson for [grade level] explaining [government concept, law, or process]. Include a real-world example and a student roleplay activity.

---

## 5. Cultural Comparison Prompt

Develop a lesson where students compare and contrast two cultures related to [topic]. Include a graphic organizer and discussion prompts.

---

## 6. Cause & Effect Prompt

Plan an activity for [grade level] that examines the causes and effects of [historical event or policy]. Include a cause-effect chart and reflection questions.

---

## 7. Historical Roleplay Prompt

Create a roleplay activity for [grade level] where students take on the perspectives of people involved in [historical event]. Include character cards and guiding questions.

---

## 8. Timeline Creation Prompt

Design a lesson for [grade level] where students create a timeline of events related to [topic]. Include at least 8 key events and short descriptions.

---

## 9. Current Events Connection Prompt

Plan a [grade level] activity where students connect a historical event to a current event. Include research questions and a group discussion component.

---

## 10. Debate Prompt

Create a structured debate for [grade level] on a historical or civic issue related to [topic]. Include debate format, guiding questions, and reflection prompts.

---

# Pro Tip:

When generating Social Studies content with AI, specify the **geographic region, time period, and desired perspective** to get accurate and relevant outputs.

# Chapter 9 – Arts, Music, and PE Prompts

## Purpose of This Chapter

Arts, music, and physical education allow students to express themselves, develop new skills, and build confidence. These prompts help teachers quickly generate AI-assisted lesson ideas that inspire creativity and promote physical wellness.

## Prompts

### 1. Visual Arts Project Prompt

Create a [grade level] art lesson on [theme or technique]. Include an introduction to the concept, a step-by-step project guide, and reflection questions.

### 2. Music Composition Prompt

Design a [grade level] music activity where students compose a short piece inspired by [theme or event]. Include basic music theory guidance and sharing instructions.

### 3. Art History Connection Prompt

Create a lesson for [grade level] that introduces [artist or movement]. Include background information, an analysis activity, and a hands-on art project.

### 4. Drama Roleplay Prompt

Plan a drama activity for [grade level] based on [historical event, novel, or theme]. Include character prompts, acting tips, and discussion questions.

## 5. PE Skill Development Prompt

Design a PE lesson for [grade level] focused on improving [specific physical skill]. Include warm-up, skill drills, and a closing activity.

## 6. Cross-Curricular Arts Integration Prompt

Create a lesson for [grade level] that combines [subject] with visual or performing arts to reinforce learning on [topic]. Include activity steps and assessment ideas.

## 7. Movement & Wellness Prompt

Develop a physical activity for [grade level] that promotes both fitness and mindfulness. Include breathing exercises, movement instructions, and a reflection prompt.

## 8. Music Appreciation Prompt

Plan a listening activity for [grade level] featuring [musical genre, composer, or song]. Include guiding questions and a creative response activity.

### 9. Seasonal Performance Prompt

Create a [grade level] performance activity tied to a seasonal or school event. Include planning steps, rehearsal tips, and audience engagement ideas.

---

### 10. Outdoor Team Game Prompt

Design a [grade level] outdoor PE game for [number] students that promotes teamwork and strategy. Include rules, variations, and safety notes.

---

## Pro Tip:

When using AI for creative subjects, provide **clear parameters**, such as time, materials, and skill level, to ensure the activity is realistic for your setting.

# Chapter 10 – Sub Plans with AI

## Purpose of This Chapter

Substitute teachers often need clear, self-contained lessons that require minimal prep and can be taught without deep subject expertise. These prompts help create AI-generated sub plans that keep students engaged while ensuring continuity in learning.

---

## Prompts

### 1. One-Day Lesson Plan Prompt

Create a full-day sub plan for [grade level] in [subject] on [topic]. Include clear instructions, student handouts, and a short assessment.

---

### 2. Emergency Sub Plan Prompt

Generate a one-period emergency lesson for [grade level] in [subject] on [topic] that requires no special materials. Include an opening activity, main task, and exit ticket.

---

### 3. Review & Practice Prompt

Plan a review lesson for [grade level] in [subject] on [unit or topic] with independent practice activities and an answer key for the sub.

---

### 4. Cross-Curricular Sub Plan Prompt

Create a [grade level] lesson that combines [subject 1] and [subject 2] for use as a sub plan. Include simple activities that reinforce both subjects.

### 5. Read & Respond Prompt

Provide a short reading passage for [grade level] in [subject] on [topic]. Include comprehension questions, a vocabulary list, and a short writing task.

### 6. Creative Project Prompt

Develop a creative project for [grade level] in [subject] on [topic] that can be started and mostly completed in one class period. Include step-by-step instructions.

### 7. Discussion-Based Lesson Prompt

Create a discussion-focused lesson for [grade level] in [subject] on [topic]. Include 5–7 open-ended questions and a small-group discussion guide.

### 8. Worksheet & Activity Pack Prompt

Generate a set of worksheets for [grade level] in [subject] on [topic]. Include answer keys and clear instructions for independent completion.

### 9. Educational Video Integration Prompt

Plan a [grade level] sub lesson in [subject] on [topic] that includes a recommended free educational video, discussion questions, and a follow-up activity.

---

### 10. Game Day Prompt

Create a [grade level] educational game for [subject] on [topic] that can be run in 30–45 minutes. Include setup instructions, rules, and a scoring system.

---

## Pro Tip:

When making sub plans with AI, be sure to **specify no special materials "** if you want to ensure substitutes can run the lesson with only basic classroom supplies.

# Chapter 11 – Flipped Classroom & Hybrid Learning Prompts

## Purpose of This Chapter

The flipped classroom model maximizes class time for discussion, practice, and collaboration by having students learn core concepts independently. These prompts help generate AI-supported lesson structures for both fully flipped and hybrid learning environments.

## Prompts

### 1. Video-Based Flipped Lesson Prompt

Create a [subject] lesson for [grade level] on [topic] where students watch a short video before class. Include a viewing guide, in-class discussion activity, and a follow-up quiz.

### 2. Reading-First Flipped Lesson Prompt

Plan a flipped lesson where students read a short article or text on [topic] at home, then complete a collaborative in-class activity to apply the concepts.

### 3. Interactive Online Module Prompt

Design an online lesson for [topic] in [subject] that students complete before class, using free interactive tools. Include clear pre-class instructions and an in-class hands-on activity.

## 4. Hybrid Lab Lesson Prompt

Create a hybrid lesson where students review safety and procedures for [lab topic] at home, then conduct the experiment in class. Include pre-lab questions and a post-lab reflection.

---

## 5. Peer Teaching Flipped Prompt

Plan a flipped activity where students learn a mini-topic independently, then teach it to peers during class time. Include preparation steps and peer evaluation forms.

---

## 6. Skill Practice & Feedback Prompt

Develop a hybrid learning sequence for [skill or concept] where students practice online at home, then receive teacher feedback in class. Include practice examples and feedback criteria.

---

## 7. Case Study Flipped Prompt

Create a case study for [topic] in [subject] that students read or watch before class. Include guiding questions for an in-class problem-solving discussion.

---

## 8. Group Project Hybrid Prompt

Design a hybrid project where research is done independently at home and group synthesis happens in class. Include milestones, collaboration tips, and deliverables.

### 9. Simulation & Roleplay Prompt

Plan a hybrid lesson where students complete a simulation online at home and roleplay scenarios in class based on their results.

### 10. Problem-Solving Challenge Prompt

Create a flipped learning activity for [topic] in [subject] where students learn key concepts at home, then solve a complex problem together in class. Include pre-class resources and in-class steps.

## Pro Tip:

When generating flipped classroom prompts with AI, always **specify time limits** for both the independent and in-class portions to maintain a realistic balance.

# Chapter 12 – Project-Based Learning Units

## Purpose of This Chapter

Project-Based Learning (PBL) engages students in extended, real-world tasks that require research, problem-solving, and teamwork. These prompts help generate AI-assisted PBL units that are standards-aligned, flexible, and highly engaging.

## Prompts

### 1. Real-World Problem-Solving Project Prompt

Create a 2-week PBL unit for [subject] in [grade level] where students solve a real-world problem related to [topic]. Include driving question, milestones, and final product guidelines.

### 2. Community Impact Project Prompt

Design a 3-week PBL unit for [grade level] where students research and propose solutions to a local community issue. Include research tasks, interviews, and presentation format.

### 3. Interdisciplinary Project Prompt

Create a PBL unit that integrates [subject 1] and [subject 2] for [grade level] on [topic]. Include learning goals, project steps, and a cross-subject final deliverable.

### 4. Historical Investigation Prompt

Plan a PBL unit for [grade level] where students investigate a historical event or figure, then present findings through a documentary or exhibit. Include research checkpoints and peer review.

---

### 5. STEM/STEAM Innovation Project Prompt

Develop a PBL unit for [grade level] where students design and build a prototype related to [science/engineering concept]. Include design process stages, testing steps, and presentation guidelines.

---

### 6. Creative Arts Integration Prompt

Design a PBL unit where students explore [theme] through art, music, or drama. Include background research, creative process steps, and a final performance or gallery walk.

---

### 7. Sustainability Challenge Prompt

Create a PBL unit for [grade level] focused on developing sustainable solutions for [environmental issue]. Include research tasks, solution design, and a public pitch presentation.

---

### 8. Cross-Cultural Exploration Prompt

Plan a PBL unit for [grade level] where students study two cultures and create a comparative multimedia project. Include research guidelines, artifact analysis, and creative output options.

### 9. Business/Entrepreneurship Project Prompt

Create a PBL unit for [grade level] where students develop a small business or social enterprise concept related to [topic]. Include market research, budgeting, and a business pitch.

### 10. Technology Integration Prompt

Design a PBL unit where students create a digital product (website, podcast, app) related to [topic]. Include technical guidance, content creation steps, and launch plan.

## Pro Tip:

When creating PBL units with AI, **request milestones and checkpoints to ensure the project remains** organized and manageable for both students and teachers.

# Chapter 13 – Seasonal & Event-Based Lessons

## Purpose of This Chapter

Seasonal and event-based lessons capture student interest by connecting learning to timely celebrations or school events. These prompts help teachers quickly create relevant, standards-aligned activities for different times of the year.

## Prompts

### 1. Holiday-Themed Lesson Prompt

Create a [subject] lesson for [grade level] on [topic] that incorporates [holiday]. Include a short historical background, themed activity, and discussion question.

### 2. Cultural Celebration Prompt

Plan a [subject] lesson for [grade level] inspired by [cultural event]. Include a hands-on activity, vocabulary list, and reflection component.

### 3. Seasonal Science Prompt

Develop a [grade level] science activity related to [season] that connects to [specific concept]. Include materials list and observation questions.

## 4. Math in the Holidays Prompt

Create a [grade level] math lesson on [topic] using data, shapes, or patterns related to [holiday or season]. Include at least 5 practice problems with solutions.

---

## 5. Literacy Event Prompt

Design a [grade level] ELA lesson inspired by [school-wide event, e.g., Read Across America]. Include a read-aloud, discussion prompts, and a short writing activity.

---

## 6. Arts & Culture Prompt

Create an arts-integrated lesson for [grade level] that celebrates [event or tradition]. Include background information, artistic creation steps, and presentation ideas.

---

## 7. Global Connections Prompt

Plan a [grade level] social studies activity comparing how [holiday or tradition] is celebrated in three different countries. Include a Venn diagram template and discussion questions.

---

## 8. School Spirit Prompt

Develop a lesson or activity for [grade level] that connects learning to a school event like Spirit Week or Field Day. Include collaborative and competitive components.

### 9. Environmental Awareness Prompt

Create a [subject] lesson for [grade level] tied to [event like Earth Day]. Include a real-world problem-solving activity and a take-home action challenge.

### 10. History of the Holiday Prompt

Design a history-focused lesson for [grade level] explaining the origins of [holiday or event]. Include a timeline activity and a creative student project.

## Pro Tip:

When using AI for seasonal or event-based lessons, **specify the cultural context and sensitivities** to ensure activities are inclusive and respectful.

# Chapter 14 – Editable Prompt Templates

## Purpose of This Chapter

These editable prompt templates provide teachers with flexible frameworks for creating AI-assisted lessons. By filling in the blanks with their subject, topic, grade level, and specific requirements, teachers can quickly generate customized content.

---

## Templates

### 1. General Lesson Plan Template

Create a [number]-minute [subject] lesson for [grade level] on [topic]. Include: learning objectives, introduction, main activities, assessment, and extension ideas.

---

### 2. Standards-Aligned Template

Develop a [subject] lesson for [grade level] on [topic] aligned with [specific standard]. Include at least two interactive activities and one formative assessment.

---

### 3. Differentiation Template

Plan a [subject] lesson on [topic] for [grade level] that includes adaptations for: struggling learners, advanced learners, and English language learners.

---

### 4. Real-World Connection Template

Create a [subject] lesson for [grade level] on [topic] that connects to a real-world issue or example. Include an engaging introduction, main task, and reflection.

---

## 5. Activity Generator Template

Develop an engaging classroom activity for [subject] on [topic] for [grade level]. Include clear instructions, needed materials, and ways to assess participation.

---

## 6. Assessment Template

Create a quiz or test for [subject] on [topic] for [grade level]. Include [number] multiple-choice questions, [number] short-answer questions, and answer keys.

---

## 7. Group Work Template

Plan a group project for [subject] on [topic] for [grade level]. Include roles, steps, collaboration tips, and grading rubric.

---

## 8. Technology Integration Template

Design a [subject] lesson for [grade level] on [topic] that uses [specific digital tool]. Include a tutorial, student activity, and assessment method.

---

## 9. Creative Project Template

Create a project-based learning activity for [subject] on [topic] for [grade level]. Include a timeline, deliverables, and presentation guidelines.

---

## 10. Seasonal/Event Template

Develop a [subject] lesson for [grade level] on [topic] that incorporates [holiday, cultural event, or seasonal theme]. Include historical background, themed activity, and assessment.

---

# Pro Tip:

Encourage teachers to **save their favorite customized prompts** in a personal "AI Prompt Bank" for quick reuse throughout the year.

# Chapter 15 – Prompt Adaptation Checklist

## Purpose of This Chapter

AI prompts are rarely perfect the first time. This checklist helps teachers refine them so the output matches classroom needs, aligns with curriculum standards, and works for all learners.

---

## Step 1 – Clarify Your Goal

- **Ask yourself:** *What exactly do I need?* (lesson plan, activity, assessment, project, etc.)
- Be clear on **grade level**, **subject**, **topic**, and **desired outcome**.

---

## Step 2 – Add Context

- Specify **state or national standards**.
- Include **time limits** (e.g., "45-minute lesson").
- Identify **classroom format** (whole group, small group, individual).

---

## Step 3 – Set Output Structure

- Request outputs in a **clear format** (e.g., numbered steps, bulleted lists, tables).
- Ask for sections (introduction, activities, assessment, extension).
- Include word or question counts where needed.

## Step 4 – Adjust for Learners

- Add differentiation requests:
    - Support for struggling learners
    - Enrichment for advanced learners
    - Language supports for English learners
- Include cultural relevance and accessibility considerations.

## Step 5 – Build in Engagement

- Request **real-world connections**.
- Include interactive elements (games, debates, roleplays).
- Add opportunities for **student voice and choice**.

## Step 6 – Double-Check Accuracy

- Review AI-generated facts, calculations, and historical details.
- Edit examples to ensure they are **age-appropriate** and **curriculum-aligned**.

## Step 7 – Make It Inclusive

- Avoid stereotypes or biased examples.
- Include diverse perspectives and voices.
- Adapt materials to be accessible for students with different abilities.

## Step 8 – Test and Tweak

- Try a small part of the AI-generated material in class.
- Gather student feedback.
- Update the prompt for future use based on what worked and what didn't.

---

**Sample Adaptation Flow:**
Original Prompt:

Create a lesson plan for Grade 8 science on ecosystems.

Adapted Prompt:

Create a 45-minute Grade 8 science lesson on ecosystems aligned to NGSS MS-LS2-3. Include:

1. An engaging introduction using a real-world example
2. A hands-on group activity with role cards
3. Differentiation for advanced and struggling learners
4. A short formative assessment with answer key

---

## Pro Tip:

Keep a **Prompt Revision Log** to track what works best for your teaching style and student needs. Over time, you'll build a personal AI playbook that's highly effective.

# Conclusion

You now have a powerful collection of AI-driven prompts, templates, and strategies to transform your lesson planning process. Whether you're creating a quick warm-up, a differentiated activity, or a multi-week project, AI can help you save time, spark creativity, and tailor instruction to meet the diverse needs of your students.

The key to success isn't just using AI, it's using it thoughtfully, ethically, and strategically. As you integrate these prompts into your planning, remember to:

- Always review AI-generated content for accuracy and appropriateness.
- Adapt prompts to reflect your teaching style, your students' needs, and your curriculum requirements.
- Use AI as a **partner**, not a replacement, in the teaching process.

By keeping your professional judgment at the center, you'll ensure AI enhances your work while protecting the integrity of your teaching.

The future of education is here. You have the tools, now it's time to experiment, adapt, and lead the way in making AI a force for positive change in the classroom.

# Additional Resources

**AI Tools for Educators:**

- **ChatGPT** – General AI text generation and lesson planning
- **MagicSchool.ai** – Education-specific AI tools and lesson plan generators
- **Canva** – Visual creation for classroom materials
- **Curipod** – Interactive lesson and slide generation
- **Diffit** – Text adaptation for different reading levels

**Professional Development Communities:**

- **ISTE (International Society for Technology in Education)** – iste.org
- **Edutopia** – edutopia.org
- **We Are Teachers** – weareteachers.com

**Recommended Reading:**

- *Artificial Intelligence in Education* by Wayne Holmes
- *The AI Classroom* by Daniel Fitzpatrick, Amanda Fox, and Brad Weinstein
- *Teach Boldly: Using Edtech for Social Good* by Jennifer Williams

**Final Encouragement:**
Please think of this playbook as your AI-powered co-teacher, ready to brainstorm, structure, and inspire whenever you need it. Continue experimenting with prompts, share your successes with other educators, and build your own personal library of AI resources.

The best lessons are still yours. AI just helps you make them faster, sharper, and more engaging.

## About The Author

**Dr. Natoshia Anderson** is the strategist you call when you want your ed tech initiatives to reach real classrooms and actually make a difference. A former engineer and national STEM curriculum designer, she has helped schools and nonprofits worldwide develop tech-integrated programs that drive equity and engagement. She's secured over $1.85 million in funding, trained thousands of educators, and developed scalable learning experiences aligned with CSTA, ISTE, and NGSS standards. As the CEO and Founder of The Anderson Strategy Group and former Director of Programs & Partnerships at STE(A)M Truck, Dr. Anderson helps ed tech companies build authentic partnerships with educators of color and underserved communities. Bring her in to talk equity by design, culturally responsive tech, or how to avoid "diversity theater."

www.ingramcontent.com/pod-product-compliance
Lightning Source LLC
LaVergne TN
LVHW051430080426
835508LV00022B/3326